Copyright © TITUS MIRACLE

Table of Contents

INTRODUCTION 3

Definition and Scope of Animal Husbandry 4

Basic Principles of Animal Nutrition 9

Livestock Breeds and Selection 15

Animal Housing and Welfare 24

Reproduction and Breeding Management 30

Health and Disease Management 36

Grazing and Forage Management 43

Dairy Farming 50

Poultry Farming 58

Swine Production 66

Sustainable and Organic Animal Husbandry . 74

Emerging Trends and Technologies 80

CONCLUSION 84

INTRODUCTION

Animal husbandry, also known as livestock farming, is an integral component of agriculture that involves the breeding, care, and management of animals for various purposes such as food, fiber, labor, and companionship.

The practice dates back thousands of years and has played a crucial role in the development and sustenance of human societies worldwide.

In this book, we delve into the multifaceted aspects of animal husbandry, exploring its definition, scope, historical evolution, and its profound importance in both agriculture and society.

Definition and Scope of Animal Husbandry

Animal husbandry encompasses a wide range of activities related to the domestication and management of animals. At its core, it involves breeding and raising livestock for diverse purposes, including the production of meat, milk, eggs, wool, and other by-products.

The scope of animal husbandry extends beyond mere economic considerations, encompassing the welfare, health, and sustainable utilization of animals.

This discipline integrates knowledge from various fields such as biology, genetics, nutrition, veterinary science, and economics. It involves not only the practical aspects of caring for animals but also the strategic planning and decision-making required for successful and ethical animal farming.

Historical Evolution of Animal Husbandry

The roots of animal husbandry trace back to the earliest human settlements, where our ancestors began domesticating wild animals for companionship and assistance in hunting and agriculture.

The shift from a nomadic lifestyle to settled agriculture marked a pivotal moment in the evolution of animal husbandry. Animals like cattle, goats, and sheep were among the first to be domesticated, serving as sources of meat, milk, and hides.

Throughout history, different civilizations have contributed to the development and refinement of animal husbandry practices. Ancient Egyptians, for example, revered cattle and used them for agricultural work. In medieval Europe, the manorial system led to the establishment of large-scale feudal farms, where animals played a crucial role in sustaining local economies.

The industrial revolution in the 18th and 19th centuries brought significant changes to animal husbandry. Advances in technology, transportation, and breeding practices led to increased efficiency and productivity. The advent of modern veterinary medicine further improved animal health and disease management, allowing for larger and healthier livestock populations.

Importance of Animal Husbandry in Agriculture and Society

Animal husbandry holds immense significance in both agricultural and societal contexts. From an agricultural perspective, it contributes substantially to food production, supplying essential commodities such as meat, milk, eggs, and wool. Livestock also provide valuable by-products like leather, manure, and draft power, supporting various agricultural activities.

Beyond the economic benefits, animal husbandry plays a crucial role in sustainable agriculture. Livestock contributes to nutrient cycling by converting crop residues and by-products into high-quality protein sources. Their grazing activities can help maintain pasture ecosystems, preventing the accumulation of dry matter and reducing the risk of wildfires.

In societal terms, animal husbandry has cultural, social, and nutritional implications. Livestock farming is deeply ingrained in the cultural practices of many communities, influencing traditions, rituals, and celebrations. Additionally, animal husbandry provides employment opportunities in rural areas, contributing to poverty alleviation and community development.

Animal husbandry also addresses global challenges such as food security and malnutrition.

Livestock products are rich sources of essential nutrients, and well-managed animal farming systems contribute to balanced diets. In developing countries, small-scale animal husbandry can be a lifeline for vulnerable populations, providing both sustenance and income.

Furthermore, animals often serve as companions and therapy aids, playing essential roles in human well-being. The bond between humans and animals has therapeutic effects, reducing stress and promoting mental health. This dual role of animals as providers of essential resources and companions underscores the holistic importance of animal husbandry in society.

In this book, we will explore the intricacies of animal husbandry, covering topics such as nutrition, breeding, health management, and sustainable practices.

Through a comprehensive examination of the principles and practices that govern animal husbandry, readers will gain insights into the dynamic and evolving nature of this field, understanding its historical roots and its pivotal role in shaping the present and future of agriculture and society.

Basic Principles of Animal Nutrition

Animal nutrition is a cornerstone of successful animal husbandry, influencing the health, productivity, and overall well-being of livestock. In this chapter, we will explore the fundamental principles of animal nutrition, focusing on the understanding of nutritional requirements for different species, the various types of feed available, and the art of formulating balanced diets to ensure optimal animal health.

Understanding Nutritional Requirements for Different Species

The nutritional needs of animals are diverse and depend on factors such as species, age, sex, reproductive status, and physiological activity. Understanding these requirements is essential for providing appropriate diets that meet the specific needs of each animal.

Different species have distinct dietary preferences and requirements. For example, ruminants like cattle have a complex stomach structure that allows them to efficiently digest fibrous plant materials, while monogastric animals like pigs and poultry require diets with higher energy density and easily digestible nutrients. Additionally, factors such as the stage of life, whether an animal is in growth, reproduction, or maintenance, influence its nutritional demands.

Nutritional requirements encompass essential nutrients, including proteins, carbohydrates, fats, vitamins, minerals, and water. Proteins are vital for growth and tissue repair, carbohydrates provide energy, fats contribute to energy storage and insulation, vitamins and minerals are essential for various physiological functions, and water is critical for hydration and nutrient transport.

Types of Feed and Their Nutritional Components

A diverse array of feeds is available to fulfill the nutritional needs of animals. These include roughages (fibrous plant materials like hay and grass), concentrates (energy-rich feeds such as grains), protein supplements, minerals, and vitamins. Each type of feed contributes specific nutritional components, and a well-balanced diet involves combining these in appropriate proportions.

Roughages are essential for providing fiber, promoting proper digestion, and preventing issues like bloat in ruminants. Concentrates, on the other hand, supply energy-dense nutrients and are particularly crucial for animals with high energy demands, such as those in intensive production systems or during periods of rapid growth.

Protein supplements, derived from sources like soybeans or fish meal, are crucial for meeting the protein requirements of animals, especially during periods of lactation, growth, or high physical activity. Minerals and vitamins play pivotal roles in bone development, reproduction, immune function, and overall health.

Understanding the nutritional composition of each type of feed is essential for formulating balanced diets. This involves considering the energy content, protein quality, and the presence of essential vitamins and minerals. Nutrient analysis and evaluation of feed quality are integral aspects of designing diets that cater to the specific needs of different animal species and production systems.

Formulating Balanced Diets for Optimal Animal Health

The art of animal nutrition lies in formulating diets that provide the right balance of nutrients to meet the specific requirements of each animal. The process involves considering the availability and cost of different feeds, as well as the nutritional demands of the animals in question.

Balanced diets are designed to optimize growth, reproduction, milk production, and overall health. Formulators take into account the digestibility of different feed ingredients, ensuring that animals can efficiently extract nutrients from their diet. The concept of "total mixed rations" is often employed, where various feed ingredients are combined in precise proportions to create a nutritionally complete diet.

In addition to meeting basic nutritional needs, formulating diets for optimal health involves addressing specific challenges, such as preventing nutrient deficiencies or excesses, managing weight and body condition, and considering the impact of diet on immune function.

The process of formulating diets is dynamic and may require adjustments based on seasonal variations, changes in feed availability, and alterations in the physiological state of the animals. Nutritional management plans should be flexible and responsive to ensure that animals receive the appropriate nutrients throughout their lifecycle.

Moreover, advancements in nutritional science and technology have led to the development of specialized feeds and supplements. These may include feed additives that enhance nutrient absorption, probiotics for gut health, and precision feeding techniques that tailor diets to the individual needs of animals within a herd or flock.

In conclusion, the second chapter of this book delves into the foundational principles of animal nutrition. By understanding the nutritional requirements of different species, the types of feed available, and the intricacies of formulating balanced diets, readers will gain insights into the critical role nutrition plays in animal husbandry.

This knowledge is essential for promoting the health, productivity, and welfare of livestock, contributing to the overall success and sustainability of animal farming systems.

Livestock Breeds and Selection

Livestock breeds are the building blocks of animal husbandry, each uniquely suited for specific purposes such as meat, milk, wool, or labor.

In this chapter, we delve into an extensive overview of common livestock breeds, explore the criteria for selecting breeds based on their intended purpose, and discuss the significance of genetic improvement and breeding strategies in shaping the characteristics of these breeds.

Overview of Common Livestock Breeds

The diversity of livestock breeds reflects centuries of selective breeding for various traits. Different breeds have evolved to thrive in specific climates, geographic regions, and production systems. A comprehensive understanding of common livestock breeds is fundamental for successful animal husbandry.

Cattle:

- **Holstein:** Renowned for high milk production, Holsteins are the most common dairy breed globally.

- **Angus:** A popular beef cattle breed known for its marbled meat and adaptability.

- **Hereford:** Recognized for its hardiness, Hereford cattle are raised for both beef and as dual-purpose animals.

Sheep:

- **Merino:** Renowned for its fine wool, the Merino breed is often prized in the textile industry.

- **Suffolk:** A meat breed known for its muscular build, well-suited for lamb production.

- **Dorset:** Used for both meat and wool, Dorset sheep are adaptable and hardy.

Goats:

- **Boer:** A meat goat breed with excellent growth rates and carcass quality.

- **Nubian:** Known for high milk production, Nubians are a popular dairy goat breed.

- **Angora:** Raised primarily for its luxurious fiber, the Angora goat produces mohair.

Pigs:

- **Duroc:** A swine breed favored for its lean meat and efficient growth.

- **Yorkshire:** Known for high-quality pork production and excellent maternal instincts.

- **Hampshire:** A dual-purpose breed, Hampshire pigs are valued for both meat and lard.

Each breed possesses unique characteristics, including size, coat color, resistance to diseases, reproductive traits, and adaptability to specific environments. The choice of breed is a critical decision in animal husbandry, as it directly influences the success of the farming enterprise and the overall productivity of the livestock.

Criteria for Selecting Breeds Based on Purpose

The selection of livestock breeds should align with the intended purpose of the farming operation. Different breeds are optimized for specific outputs, whether it be high milk yield, superior wool quality, or efficient meat production. Selecting breeds based on purpose involves considering a variety of criteria:

1. Meat Production:

    *   **Growth Rate:** Breeds like Boer goats and Duroc pigs are renowned for their rapid growth, making them ideal for meat production.

    * **Carcass Quality:** Animals with desirable meat characteristics, such as leanness and marbling, are prioritized for meat production.

2. Milk Production:

* **Milk Yield:** Dairy breeds like Holsteins and Nubians are selected for their high milk production, critical for dairy farming.

* **Butterfat Content:** Some breeds, like Jerseys, are known for their high butterfat content, ideal for cheese and butter production.

3. Wool Production:

* **Fiber Quality:** Merino sheep are valued for their fine wool, crucial for the textile industry.

* **Fleece Quantity:** Breeds like Romney sheep are selected for their abundant fleece, providing a balance between quality and quantity.

4. Dual-Purpose:

* **Adaptability:** Certain breeds, such as Hampshire pigs and Dorset sheep, are versatile and can serve dual purposes, providing both meat and other valuable by-products.

*** Efficiency:** Dual-purpose breeds are often selected for their efficiency in resource utilization, making them suitable for a range of production systems.

Genetic Improvement and Breeding Strategies

Genetic improvement is a continuous process aimed at enhancing desirable traits within a population. In animal husbandry, selective breeding is a cornerstone of genetic improvement, influencing traits like productivity, disease resistance, and adaptability. Understanding breeding strategies is pivotal for achieving long-term goals in livestock farming.

1. Selective Breeding:

*** Traits Selection:** Farmers selectively breed animals with desirable traits, such as high milk yield, superior wool quality, or efficient feed conversion.

*** Breeding Value:** The genetic potential of animals is assessed, and those with higher breeding values are chosen as parents to pass on favorable traits to the next generation.

2. Crossbreeding:

* **Hybrid Vigor:** Crossbreeding involves mating animals of different breeds to capitalize on hybrid vigor, resulting in offspring with improved performance and resilience.

* **Complementary Traits:** Breeds with complementary traits are chosen for crossbreeding, combining the strengths of each to achieve a more balanced and productive animal.

3. Inbreeding and Linebreeding:

* **Genetic Diversity:** Controlled mating within closely related individuals (linebreeding) or more closely related animals (inbreeding) can help fix desirable traits but should be approached cautiously to avoid negative effects on genetic diversity.

* **Pedigree Analysis:** Detailed pedigree analysis is essential to understand the genetic relationships between animals and make informed breeding decisions.

4. Biotechnology and Genomics:

* **Marker-Assisted Selection:** Advances in biotechnology allow for the identification of specific genetic markers associated with desirable traits, enabling more precise breeding.

* **Genomic Selection:** The use of genomics in breeding programs allows for a more comprehensive understanding of an animal's genetic potential, improving the accuracy of trait predictions.

Genetic improvement is a dynamic process that requires careful planning, documentation, and adaptability to changing environmental and market conditions. Breeding strategies should be aligned with the goals of the farm, considering both short-term gains and long-term sustainability.

By providing an overview of common livestock breeds, discussing criteria for breed selection based on specific purposes, and delving into the intricacies of genetic improvement and breeding strategies, readers gain a holistic understanding of the foundational principles that govern successful livestock farming. The choices made in selecting and breeding animals profoundly impact the productivity, resilience, and sustainability of animal husbandry operations, making this knowledge essential for anyone engaged in or aspiring to enter the field.

Animal Housing and Welfare

Is a comprehensive field that encompasses the care, breeding, and management of domesticated animals. Within this discipline, one crucial aspect is the design and construction of animal shelters, as well as the promotion of animal welfare. In this discussion, we will delve into the intricacies of animal housing and welfare, exploring the design and construction of shelters, factors influencing animal welfare, and best practices for creating a comfortable and safe environment for animals.

Design and Construction of Animal Shelters:

1. Purpose and Functionality:

Animal shelters serve multiple purposes, ranging from protecting adverse weather conditions to offering a safe space for animals to engage in natural behaviors. The design must consider the specific needs of the animals, accounting for species, size, and behavior.

2. Materials and Construction:

Selecting appropriate materials is paramount to the durability and functionality of animal shelters. Weather-resistant materials, proper insulation, and adequate ventilation are key considerations. The construction should ensure structural integrity while allowing for easy cleaning and maintenance.

3. Space Allocation:

The layout and size of animal shelters should provide sufficient space for movement, rest, and social interactions. Adequate space allocation helps prevent overcrowding, which can lead to stress and aggressive behavior among animals.

4. Environmental Control:

Animal shelters must incorporate features that allow for climate control. This may include heating, cooling, and ventilation systems to maintain optimal conditions, especially in extreme weather. Adequate lighting is also essential for promoting natural behaviors.

Factors Influencing Animal Welfare:

1. Physical Health:

The overall health of animals is a crucial factor in welfare. Access to clean water, balanced nutrition, and veterinary care are fundamental to ensuring the physical well-being of animals. Regular health checks and preventive measures help in maintaining a healthy population.

2. Behavioral Considerations:

Understanding and accommodating the natural behaviors of animals is essential for promoting their welfare. Enrichment activities, appropriate socialization, and play opportunities contribute to the psychological well-being of animals.

3. Stress Management:

Stress can significantly impact animal welfare. Factors such as noise levels, handling practices, and the social environment can induce stress in animals. Implementing stress-reducing measures, such as quiet zones and proper handling techniques, is crucial.

4. Disease Prevention:

Disease prevention is a critical aspect of animal welfare. Proper sanitation, quarantine protocols, and vaccination programs are essential to control the spread of diseases within animal populations.

Best Practices for Creating a Comfortable and Safe Environment:

1. Routine Inspections:

Regular inspections of animal shelters are necessary to identify and address any structural issues, maintenance requirements, or potential hazards. This ensures that the environment remains safe and comfortable for the animals.

2. Environmental Enrichment:

Incorporating environmental enrichment activities, such as toys, structures, and varied terrain, helps stimulate natural behaviors, reducing boredom and promoting mental well-being.

3. Training and Handling Protocols:

Proper training and handling of animals contribute significantly to their welfare. Educating caretakers on appropriate handling techniques and implementing positive reinforcement methods create a stress-free environment.

4. Record Keeping:

Maintaining detailed records of the health, behavior, and living conditions of animals allows for better monitoring and assessment of their welfare. This information is valuable in identifying patterns or issues that may need attention.

5. Community and Stakeholder Involvement:

Involving the community and stakeholders in animal husbandry practices fosters a collaborative approach to animal welfare. Education and awareness programs can create a shared responsibility for the well-being of animals.

A holistic approach that considers physical health, behavioral needs, and environmental factors is essential for creating a comfortable and safe environment for domesticated animals.

By incorporating best practices and engaging in continuous improvement, we can ensure the well-being of animals under our care, contributing to a sustainable and ethical approach to animal husbandry.

Reproduction and Breeding Management

Reproduction and breeding management are critical aspects of animal husbandry, playing a pivotal role in maintaining and improving livestock populations. The book under discussion provides a comprehensive exploration of these topics, covering reproductive anatomy and physiology, estrus detection, breeding techniques, and pregnancy management, as well as the care of newborns. Let's delve into the intricate details of each of these components.

 Reproductive Anatomy and Physiology:

1. Male Reproductive System:

The male reproductive system comprises organs such as the testes, epididymis, vas deferens, and accessory glands. Understanding the anatomy and function of these structures is fundamental for assessing male fertility. The book elucidates the roles of hormones in regulating sperm production and the factors affecting sperm quality.

2. Female Reproductive System:

The female reproductive system involves a complex interplay of organs, including the ovaries, fallopian tubes, uterus, and vagina. The book provides a detailed analysis of the menstrual cycle, ovulation, and the hormonal fluctuations that influence the reproductive process.

3. Endocrine Regulation:

Hormones, such as follicle-stimulating hormone (FSH), luteinizing hormone (LH), progesterone, and estrogen, play crucial roles in regulating reproductive processes. The book explores how these hormones orchestrate the various stages of the reproductive cycle, from follicular development to gestation.

Estrus Detection and Breeding Techniques:

1. Estrus Detection:

Estrus, or the heat cycle, is a pivotal period in the reproductive cycle of females. Proper detection of estrus is crucial for successful breeding. The book outlines behavioral and physical signs of estrus, emphasizing the importance of keen observation and advanced technologies, such as electronic heat detection devices.

2. Artificial Insemination (AI):

Artificial insemination is a widely used breeding technique in animal husbandry. The book discusses the principles of AI, including semen collection, processing, and insemination procedures. It explores the advantages of AI in terms of genetic improvement, disease control, and overall reproductive efficiency.

3. Embryo Transfer:

Embryo transfer is an advanced reproductive technology that allows for the transfer of embryos from a genetically superior donor to multiple recipients. The book delves into the techniques involved in embryo collection, preservation, and transfer, as well as the potential applications and benefits of this method.

4. Genetic Selection Strategies:

The book addresses the importance of genetic selection in breeding programs. It elucidates the criteria for selecting superior breeding stock based on traits such as fertility, milk production, meat quality, and disease resistance. The integration of modern technologies, like genomics, in genetic selection is also explored.

Pregnancy Management and Care of Newborns:

1. Pregnancy Diagnosis:

Early and accurate pregnancy diagnosis is crucial for effective management. The book discusses various diagnostic methods, including palpation, ultrasound, and hormonal assays. It emphasizes the importance of timely identification of pregnant animals for appropriate nutritional and healthcare interventions.

2. Nutritional Requirements During Pregnancy:

The nutritional needs of pregnant animals significantly impact the health of both the mother and the developing fetus. The book provides guidelines on formulating balanced diets that meet the increased energy and nutrient demands during gestation, ensuring optimal fetal development.

3. Parturition and Dystocia:

The book delves into the physiological processes of parturition (birth) and the potential complications associated with dystocia (difficult birth). It explores preventive measures, such as proper nutrition and monitoring, to minimize the risks of dystocia, as well as intervention techniques when assistance is required.

4. Neonatal Care:

The care of newborns is a critical aspect of ensuring their health and survival. The book addresses topics like colostrum management, vaccination protocols, and appropriate housing for neonates. It emphasizes the importance of early bonding and proper hygiene to minimize the risk of infections.

Reproduction and Breeding Management provides a thorough exploration of the intricate processes involved in the reproductive cycle of domesticated animals.

From understanding the anatomy and physiology of both male and female reproductive systems to implementing advanced breeding techniques and ensuring proper care during pregnancy and neonatal stages, the book offers a comprehensive guide for practitioners in the field of animal husbandry.

By integrating scientific principles, technological advancements, and practical management strategies, this resource serves as an invaluable tool for those dedicated to the improvement and sustainability of livestock populations.

Health and Disease Management

Health and disease management in animal husbandry is a critical aspect of ensuring the well-being and productivity of livestock.

The book under consideration provides a comprehensive guide to practitioners, covering preventive healthcare practices, common diseases in livestock and their control, as well as vaccination schedules and disease monitoring. In this detailed exploration, we will delve into the key aspects covered in the book.

Preventive Healthcare Practices:

1. Biosecurity Measures:

The book emphasizes the importance of implementing biosecurity measures to prevent the introduction and spread of diseases within livestock populations.

This includes strict quarantine protocols for new animals, control of visitors and equipment, and the establishment of secure and controlled environments to minimize disease transmission.

2. Nutritional Management:

Proper nutrition is fundamental to maintaining the overall health and resilience of animals. The book discusses balanced feed formulations, appropriate grazing practices, and the importance of providing essential nutrients to prevent nutritional deficiencies. A focus on nutrition contributes to a robust immune system and disease resistance.

3. Herd Health Programs:

Implementing herd health programs involves regular health check-ups, monitoring for signs of illness, and timely intervention. The book provides guidelines on designing and implementing effective herd health plans, including routine vaccinations, parasite control, and nutritional supplementation.

4. Environmental Management:

Livestock health is closely linked to the environment in which they are raised. Proper waste management, ventilation, and housing conditions are crucial for preventing diseases related to stress, overcrowding, and unsanitary living conditions. The book offers insights into creating optimal environmental conditions for animal health.

Common Diseases in Livestock and Their Control:

1. Parasitic Infections:

Livestock is susceptible to various parasites, including internal worms and external pests. The book delves into the identification, treatment, and prevention of parasitic infections. It discusses deworming protocols, pasture management, and the use of anti-parasitic medications to control infestations.

2. Respiratory Diseases:

Respiratory infections are common in livestock, especially in crowded or poorly ventilated conditions. The book provides information on recognizing symptoms of respiratory diseases, implementing vaccination strategies, and improving ventilation to minimize the risk of outbreaks.

3. Reproductive Disorders:

Disorders affecting the reproductive system can have significant implications for livestock productivity. The book explores the causes and control measures for reproductive disorders, including infertility, abortions, and postpartum complications. It emphasizes the importance of maintaining optimal reproductive health for breeding success.

4. Metabolic Diseases:

Metabolic disorders, such as ketosis, acidosis, and milk fever, can impact the overall health and productivity of livestock. The book discusses preventive measures, including proper nutrition, monitoring for early signs of metabolic imbalances, and implementing management practices to mitigate the risk of these conditions.

Vaccination Schedules and Disease Monitoring:

1. Vaccination Protocols:

The book outlines comprehensive vaccination schedules for different livestock species. It discusses the types of vaccines available, their modes of administration, and the importance of timing in providing immunity.

Vaccination is a key component of disease prevention, and the book provides guidelines for developing tailored vaccination programs based on regional disease prevalence and specific herd conditions.

2. Disease Monitoring and Surveillance:

A proactive approach to disease management involves continuous monitoring and surveillance. The book details methods for disease surveillance, including regular health checks, diagnostic testing, and the use of technology for early detection of diseases. Monitoring allows for timely intervention and the implementation of control measures to prevent the spread of infectious agents.

3. Quarantine Procedures:

The book emphasizes the importance of quarantine procedures for introducing new animals into a herd or flock. It discusses the duration of quarantine, health assessments, and testing to ensure that new animals do not bring infectious diseases into the existing population. Strict quarantine measures are crucial for preventing disease outbreaks.

4. Integrated Disease Management:

Integrating various disease management strategies is key to a holistic approach. The book encourages the use of integrated pest management, combining vaccination, nutrition, biosecurity, and regular monitoring to create a robust defense against diseases. **This approach enhances the overall health and resilience of livestock.**

Health and Disease Management serves as a comprehensive resource for practitioners in the field. By addressing preventive healthcare practices, common diseases, and vaccination schedules, the book equips livestock managers with the knowledge and tools needed to maintain a healthy and productive animal population.

Emphasizing the importance of biosecurity, nutrition, environmental management, and integrated disease control, the book promotes a holistic approach to animal health. As the field of animal husbandry continues to evolve, this resource contributes to the advancement of practices that prioritize the well-being and sustainability of livestock populations.

Grazing and Forage Management

Grazing and forage management play a pivotal role in the sustenance and productivity of livestock. The book under consideration delves into the intricacies of sustainable grazing practices, forage crop selection and management, and rotational grazing systems.

This comprehensive guide equips practitioners in animal husbandry with knowledge and strategies to optimize grazing resources, enhance forage quality, and promote sustainable land management.

Sustainable Grazing Practices:

1. Holistic Resource Management:

The book emphasizes a holistic approach to grazing management, recognizing the interconnectedness of soil health, vegetation, and animal well-being. Sustainable grazing practices involve understanding the carrying capacity of the land, implementing rest periods to allow forage regrowth, and incorporating conservation measures to prevent soil degradation.

2. Grazing Intensity and Duration:

Proper grazing intensity and duration are critical components of sustainable practices. The book provides guidelines on monitoring and adjusting stocking rates to prevent overgrazing, which can lead to soil erosion and depletion of forage resources. It advocates for rotational grazing systems that allow for strategic rest periods, promoting regrowth and maintaining pasture health.

3. Biodiversity Conservation:

Sustainable grazing practices aim to preserve and enhance biodiversity on grazing lands. The book discusses the importance of maintaining a diverse range of plant species to support a healthy ecosystem. Encouraging native grasses and forbs contributes to soil stability, enhances nutrient cycling, and provides a balanced diet for livestock.

4. Water Management:

Adequate water availability is crucial for both livestock and forage health. The book provides insights into water management strategies, including the placement of water sources, proper fencing to control access and efficient irrigation practices. Well-managed water resources contribute to improved forage utilization and overall sustainability.

Forage Crop Selection and Management:

1. Forage Species Selection:

The book explores the process of selecting suitable forage species based on climate, soil type, and livestock nutritional requirements. It provides information on both cool-season and warm-season forages, emphasizing the importance of diversity to meet the varying needs of livestock throughout the year.

2. Forage Establishment and Maintenance:

Successful forage management begins with proper establishment. The book covers topics such as seedbed preparation, planting techniques, and fertility management. It delves into strategies for maintaining forage stands, including weed control, fertilization, and disease prevention, to ensure the sustained productivity of grazing lands.

3. Forage Quality and Nutritional Content:

The nutritional content of forage directly impacts the health and productivity of livestock. The book provides insights into assessing forage quality, including methods for testing protein, fiber, and mineral content. Understanding these aspects allows for informed decision-making in terms of supplementation and overall animal nutrition.

4. Seasonal Forage Management:

Recognizing the seasonality of forage growth is essential for effective management. The book discusses strategies for adjusting grazing patterns and forage utilization based on seasonal changes. This includes planning for periods of dormancy, optimizing forage production during peak growing seasons, and ensuring adequate winter forage reserves.

Rotational Grazing Systems:

1. Principles of Rotational Grazing:

Rotational grazing involves dividing pastures into smaller paddocks and systematically rotating livestock through these areas. The book elucidates the principles of rotational grazing, emphasizing its benefits in terms of forage utilization, soil health, and livestock performance. It guides determining appropriate stocking densities and rotation intervals.

2. Improved Forage Utilization:

Rotational grazing enhances forage utilization by preventing selective grazing and promoting uniform forage consumption. The book discusses how controlled access to paddocks and strategic rotation prevent overgrazing, allowing for more efficient forage use. This results in improved nutrition for livestock and increased forage productivity.

3. Soil Health and Fertility:

The book explores the positive impact of rotational grazing on soil health and fertility. By providing periods of rest for grazed areas, the soil has the opportunity to recover, reducing compaction and promoting nutrient cycling. This sustainable approach contributes to long-term soil fertility and pasture resilience.

4. Disease and Parasite Control:

Rotational grazing can aid in disease and parasite control. By moving livestock through different paddocks, the book explains how parasites are less likely to build up in the environment. This reduces the need for chemical interventions and promotes a healthier living environment for the animals.

Grazing and Forage Management offers a comprehensive guide to practitioners seeking to optimize the use of grazing resources, enhance forage quality, and promote sustainable land management.

By exploring sustainable grazing practices, forage crop selection and management, and rotational grazing systems, the book addresses the interconnected challenges of maintaining healthy pastures, preserving biodiversity, and ensuring the well-being of livestock.

Practitioners in animal husbandry can benefit from the practical insights and scientific principles outlined in the book, applying them to real-world scenarios. Ultimately, the adoption of sustainable grazing practices contributes not only to the productivity of livestock operations but also to the long-term health and resilience of grazing lands.

As the field of animal husbandry continues to evolve, the knowledge presented in this book serves as a valuable resource for those dedicated to sustainable and efficient forage and grazing management.

Dairy Farming

Dairy farming is a specialized branch of animal husbandry dedicated to the management and production of milk from dairy cattle.

The book under consideration provides an in-depth exploration of key aspects of dairy farming, including dairy cattle breeds and their characteristics, milk production and quality control, as well as dairy farm management and technology.

Dairy Cattle Breeds and Their Characteristics:

1. Holstein:

Holsteins are one of the most prevalent dairy cattle breeds worldwide, renowned for their high milk production. Recognized by their distinctive black-and-white color pattern, Holsteins are large-framed cows with excellent feed efficiency. They excel in converting feed into milk, making them a popular choice for commercial dairy operations.

2. Jersey:

Jersey cattle are known for their fawn color and gentle disposition. Despite their smaller size compared to Holsteins, Jerseys are valued for the high butterfat content in their milk. This breed is adaptable to various climates and is often chosen by dairy farmers looking to produce premium-quality dairy products.

3. Guernsey:

Guernseys are medium-sized cows with a reddish-brown and white coat. They are recognized for the rich flavor and high butterfat and protein content in their milk. Guernseys are adaptable and known for their longevity, making them an appealing choice for sustainable and small-scale dairy farming.

4. Ayrshire:

Ayrshires are hardy cows with red and white coats. Known for their versatility, Ayrshires are suitable for both grazing systems and confinement operations. They produce milk with a balanced composition of fat and protein. Their adaptability makes them valuable in various farming environments.

5. Brown Swiss:

Brown Swiss cattle are characterized by their solid brown color and large, sturdy build. They are known for producing high volumes of milk with a balanced composition. Brown Swiss cows are particularly valued for their durability, making them well-suited to different management systems and environmental conditions.

Milk Production and Quality Control:

1. Nutritional Management:

The book emphasizes the critical role of nutrition in dairy cattle management. Proper nutrition is essential for maximizing milk production and ensuring the quality of the milk produced. It delves into the formulation of balanced diets, addressing the specific nutrient requirements of lactating cows to support optimal milk yield and composition.

2. Reproductive Management:

Efficient reproductive management is crucial for maintaining a consistent and profitable milk production cycle. The book discusses breeding strategies, artificial insemination techniques, and reproductive health monitoring. Timely and successful breeding contributes to maintaining a productive herd and achieving desired genetic traits.

3. Milk Quality and Hygiene:

Ensuring the quality and safety of milk is paramount in dairy farming. The book provides guidelines for maintaining strict hygiene practices during milking, proper sanitation of milking equipment, and routine testing for contaminants. It explores the implementation of quality control measures to meet regulatory standards and consumer expectations.

4. Milking Technology:

Technological advancements in milking equipment play a crucial role in modern dairy farming. The book covers the use of milking machines, automated systems, and monitoring devices. It details best practices for milking procedures to minimize stress on the animals and optimize milk extraction efficiency.

Dairy Farm Management and Technology:

1. Herd Health Management:

The book underscores the significance of herd health management in dairy farming. It explores preventive healthcare practices, vaccination protocols, and disease monitoring strategies.

A proactive approach to herd health contributes to reducing the incidence of diseases, enhancing milk production, and ensuring the overall well-being of the herd.

2. Facility Design and Layout:

Efficient facility design is critical for the well-being of dairy cattle and the ease of farm management. The book discusses considerations for barn design, milking parlor layout, and handling facilities. Adequate space, ventilation, and comfort for the cows are integral components of well-designed dairy facilities.

3. Precision Agriculture:

Precision agriculture technologies have revolutionized dairy farming. The book explores the use of sensors, data analytics, and automation in managing herd health, optimizing feed efficiency, and monitoring milk production. Precision agriculture enhances decision-making processes, leading to improved productivity and resource utilization.

4. Environmental Sustainability:

Sustainable dairy farming practices are increasingly important in the context of environmental stewardship. The book delves into strategies for waste management, nutrient recycling, and reducing the environmental impact of dairy operations. Implementing sustainable practices not only benefits the environment but also aligns with consumer preferences for ethically produced dairy products.

Dairy Farming provides a comprehensive guide to the intricate world of managing dairy cattle for milk production. By examining dairy cattle breeds and their characteristics, exploring milk production and quality control measures, and delving into dairy farm management and technology, the book equips practitioners with the knowledge and strategies needed for success in the dairy industry.

Dairy farming is a dynamic and evolving field, influenced by advancements in technology, changing consumer preferences, and a growing emphasis on sustainability. The information presented in this book reflects the integration of traditional practices with modern approaches, emphasizing the importance of animal welfare, milk quality, and efficient farm management.

Practitioners in dairy farming, whether experienced or new to the field, can benefit from the practical insights, scientific principles, and technological advancements discussed in the book.

As the dairy industry continues to adapt to meet global demand for high-quality dairy products, this resource serves as a valuable tool for those dedicated to the responsible and profitable management of dairy cattle.

Poultry Farming

Poultry farming is a multifaceted branch of animal husbandry that involves the breeding and rearing of domesticated birds for various purposes, including egg production and meat (broiler farming). The book under consideration comprehensively covers the diverse types of poultry, housing and equipment requirements, and the intricacies of egg production and broiler farming.

Types of Poultry:

1. Chickens:

Chickens are the most common and widely raised poultry species. They are valued for their versatility, providing eggs, meat, and feathers. The book details various chicken breeds, each with distinct characteristics such as egg-laying capacity, meat quality, and adaptability to different climates. Popular breeds include Leghorns for egg production and Cornish Cross for meat (broilers).

2. Ducks:

Ducks are another significant poultry species, appreciated for their eggs, meat, and pest control abilities. The book explores different duck breeds, such as Pekin and Khaki Campbell, known for their high egg production. Ducks are hardy birds and can thrive in a variety of environments, making them suitable for diverse poultry farming operations.

3. Turkeys:

Turkeys are primarily raised for their meat, particularly during festive seasons. The book discusses turkey breeds, including Broad Breasted White and Heritage breeds. Turkeys require specific care and nutrition to promote healthy growth and are often raised in specialized facilities to meet market demands for Thanksgiving and other occasions.

4. Quail:

Quail farming has gained popularity due to the small size of these birds and their ability to adapt to limited space. The book addresses various quail breeds, such as Coturnix quail, known for their prolific egg production. Quail farming is suitable for small-scale operations and is valued for its efficiency in egg production.

Poultry Housing and Equipment:

1. Brooding Facilities:

Brooding is a crucial stage in poultry farming, especially for chicks. The book details the design and management of brooding facilities, including brooder heaters, bedding material, and temperature control. Proper brooding ensures the health and development of chicks in their initial weeks.

2. Layer Housing:

For egg-laying poultry, appropriate layer housing is essential. The book explores various types of layer housing systems, such as cage systems, free-range systems, and aviaries. Each system has its advantages and considerations, impacting factors like egg production, bird health, and overall management efficiency.

3. Broiler Houses:

Broiler farming requires specialized housing to accommodate the rapid growth of meat-producing birds. The book discusses broiler house design, ventilation systems, and litter management. Optimizing these factors contributes to the well-being of broilers and ensures efficient meat production.

4. Equipment and Automation:

Modern poultry farming often incorporates advanced equipment and automation. The book covers essential equipment, including feeders, waterers, and lighting systems. Automation in poultry farming enhances efficiency, reduces labor requirements, and enables precise control over environmental conditions.

5. Biosecurity Measures:

The book emphasizes the importance of biosecurity in poultry farming to prevent the introduction and spread of diseases. Biosecurity measures include controlled access, sanitation protocols, and proper waste management. Implementing biosecurity practices is crucial for maintaining the health and productivity of poultry flocks.

Egg Production and Broiler Farming:

1. Egg Production:

Egg production in poultry farming involves managing laying hens for optimal egg yield and quality. The book covers aspects such as nutrition, lighting programs, and disease control to ensure consistent egg production. Different management practices are discussed, including the use of layer diets, vaccination programs, and monitoring of egg quality parameters.

2. Broiler Farming:

Broiler farming focuses on efficiently raising poultry for meat production. The book details broiler management practices, including feeding programs, growth monitoring, and disease prevention. Broilers are typically raised in controlled environments with access to specialized feeds to achieve rapid growth and high-quality meat.

3. Feed Formulation:

The book delves into the formulation of poultry feeds to meet the specific nutritional requirements of different poultry types. Feed composition varies based on factors such as age, purpose (egg production or meat), and breed. Proper feed formulation is crucial for promoting growth, health, and productivity in poultry.

4. Disease Prevention and Control:

Disease management is a significant aspect of poultry farming. The book provides insights into disease prevention through vaccination programs, biosecurity measures, and routine health monitoring. Early detection and appropriate intervention are critical for minimizing the impact of diseases on both egg-laying and broiler poultry operations.

5. Market Considerations:

The book addresses market considerations for both egg and broiler production. Understanding market demands, pricing trends, and consumer preferences is essential for planning production cycles and optimizing profitability.

Market-driven strategies may include adjusting flock sizes, diversifying product offerings, and maintaining product quality.

Poultry Farming provides an extensive exploration of the diverse world of poultry, covering various poultry types, housing and equipment requirements, and the intricacies of egg production and broiler farming. Poultry farming is a dynamic and evolving industry, influenced by technological advancements, market dynamics, and a growing emphasis on sustainability.

Practitioners in poultry farming, from small-scale producers to large commercial operations, can benefit from the practical insights, scientific principles, and management strategies discussed in the book.

 The integration of traditional practices with modern approaches, such as advanced housing systems and automation, reflects the adaptability and resilience of poultry farming.

As the demand for poultry products continues to rise globally, the information presented in this book serves as a valuable resource for those dedicated to efficient, ethical, and sustainable poultry farming practices.

From optimizing production systems to ensuring the welfare of birds, the book contributes to the ongoing advancement of poultry farming as a vital component of the broader field of animal.

Swine Production

Swine production, a crucial aspect of animal husbandry, involves the breeding, raising, and management of domestic pigs for various purposes, such as meat production and research. This multifaceted field encompasses a range of considerations, from selecting suitable breeds to ensuring optimal housing and management practices.

Swine Breeds and Characteristics:

Swine breeds play a pivotal role in determining the success of a swine production venture. Several breeds are commonly raised for pork production worldwide, each with its unique set of characteristics and advantages.

- **Yorkshire:** Known for their prolificacy, Yorkshire pigs are prized for their excellent mothering abilities. They exhibit a white coat, which makes them well-suited for various climates and environments.

- **Duroc:** Duoc pigs are recognized for their rapid growth and efficient feed conversion. With a distinctive reddish-brown coat and well-marbled meat, Durocs are often favored in pork production for their high-quality carcasses.

- **Hampshire:** Hampshire pigs are easily identifiable by their black coat with a white belt encircling their shoulders. They are valued for their meat quality, high growth rates, and adaptability to different management systems.

- **Landrace:** Landrace pigs are characterized by their long bodies and large, floppy ears. They are known for their high fertility rates, making them ideal for reproductive purposes within a swine operation.

- **Berkshire:** Berkshire pigs are prized for their flavorful and tender meat. They have a distinct black coat with white markings on their face, legs, and tail. Berkshire pork is often sought after by chefs for its superior taste and marbling.

Understanding the unique characteristics of each breed is essential for successful swine production. Factors such as growth rate, feed efficiency, reproductive performance, and adaptability to specific environments should be carefully considered when selecting breeds for a swine operation.

Swine Housing and Management:

Creating a suitable environment for swine is critical to their health, well-being, and overall productivity. Swine housing and management practices should prioritize factors such as ventilation, temperature control, and space utilization.

1. Housing Systems:

- **Farrowing Barns:** Specifically designed for sows during the farrowing (birthing) process, these barns provide a controlled environment to ensure the health and safety of both the sow and piglets.

- **Nursery Barns:** These facilities cater to the needs of piglets after weaning until they reach a suitable size to be moved to grower or finisher barns.

- **Grower-Finisher Barns:** Designed for the growing and finishing stages of pig production, these barns accommodate pigs until they reach market weight.

2. Ventilation and Temperature Control:

- **Mechanical Ventilation Systems:** Employing fans and ventilation systems helps regulate temperature and air quality within the barns.

- **Temperature-Controlled Rooms:** Ensuring a consistent temperature is crucial, especially for piglets and during extreme weather conditions.

3. Space and Environmental Enrichment:

Adequate Space: Providing sufficient space for each pig is essential for their comfort and prevents stress-related issues.

Enrichment Items: Incorporating objects such as toys or rooting materials helps stimulate natural behaviors, promoting overall well-being.

4. **Biosecurity Measures:**

- **Quarantine Protocols:** Implementing strict quarantine measures for incoming pigs prevents the introduction of diseases.

- **Sanitation Practices:** Regular cleaning and disinfection of facilities reduce the risk of disease transmission.

Pork Production and Processing:

Once the swine is raised to market weight, the focus shifts to pork production and processing, ensuring the delivery of high-quality pork products to consumers.

1. **Slaughter and Processing:**

- **Humane Slaughter Practices:** Adhering to ethical and humane slaughter practices is crucial for both animal welfare and meat quality.

- **Processing Facilities:** Modern processing facilities are equipped with advanced technology to ensure efficient and hygienic processing.

2. Quality Control and Inspection:

Quality Assurance Programs: Implementing quality assurance programs throughout the production and processing chain ensures the consistency and safety of pork products.

Government Inspection: Rigorous inspection processes by government agencies further guarantee the safety and quality of pork for consumers.

3. Value-Added Processing:

- **Cured and Smoked Products:** Processing techniques such as curing and smoking add value to pork products, creating a diverse range of options for consumers.

- **Sausage and Specialty Products:** Innovation in processing allows for the creation of various sausage and specialty pork products, meeting consumer preferences and demands.

4. Packaging and Distribution:

- **Vacuum Packaging:** Utilizing vacuum packaging helps extend the shelf life of pork products by minimizing exposure to air.

- **Cold Chain Management:** Maintaining a cold chain throughout distribution ensures that pork products reach consumers in optimal condition.

Understanding the intricacies of swine production and pork processing is essential for anyone involved in the animal husbandry industry.

From selecting the right breeds to implementing effective housing and management practices, and finally, ensuring the production of high-quality pork products, a comprehensive approach is necessary for success in this dynamic field. As the global demand for pork continues to rise, responsible and efficient swine production practices play a vital role in meeting the needs of consumers while prioritizing the welfare of the animals involved.

Sustainable and Organic Animal Husbandry

In recent years, there has been a growing emphasis on sustainable and organic practices within the field of animal husbandry. This shift is driven by concerns about environmental impact, animal welfare, and the desire for healthier, ethically produced animal products. This book, "Animal Husbandry: A Comprehensive Exploration of Sustainable and Organic Practices," delves into the principles, certification processes, and successful case studies of sustainable and organic animal husbandry.

Principles of Sustainable and Organic Practices:

1. Environmental Stewardship:

Sustainable and organic animal husbandry prioritizes environmental sustainability. This involves practices that minimize the negative impact on ecosystems, such as rotational grazing to prevent overgrazing, efficient waste management, and the use of renewable energy sources.

2. Animal Welfare:

A central tenet of these practices is ensuring the well-being of animals. This includes providing ample space for movement, access to natural light, and avoiding the use of hormones and antibiotics unless necessary for the health of the animals.

3. Biodiversity Conservation:

Sustainable and organic animal husbandry integrates with broader ecosystems, promoting biodiversity. This can involve maintaining natural habitats, avoiding monoculture practices, and incorporating diverse plant and animal species within the farm.

4. Closed-Loop Systems:

Embracing closed-loop systems is crucial for sustainability. This involves recycling and reusing resources within the farm, such as using animal manure as fertilizer, thus reducing reliance on external inputs.

Certification and Standards for Organic Animal Products:

1. Organic Certification:

To label animal products as organic, producers must adhere to strict standards set by certification bodies. These standards typically regulate feed quality, animal living conditions, and the limited use of synthetic substances. Compliance with these standards ensures that the products meet the criteria for organic certification.

2. Regulation of Feed:

Organic animal husbandry requires the use of organic feed, free from synthetic pesticides, herbicides, and genetically modified organisms (GMOs). This not only promotes the health of the animals but also aligns with the broader principles of organic farming.

3. Animal Health and Medications:

The use of antibiotics and hormones is restricted in organic animal husbandry. Producers must prioritize natural and preventive methods for animal health, with medications being a last resort. This limitation ensures the production of clean and ethically raised animal products.

4. Traceability and Transparency:

Organic certification also demands rigorous record-keeping and traceability. This allows consumers to have confidence in the origin and production methods of the animal products they purchase. Transparency is a key element in building trust between producers and consumers.

Case Studies of Successful Sustainable Animal Husbandry Operations:

1. Polyface Farm, Virginia, USA:

Polyface Farm, led by Joel Salatin, is a renowned example of sustainable animal husbandry. The farm employs rotational grazing, allowing livestock to mimic natural herd behaviors while promoting soil health. The integration of different animal species on the farm creates a balanced ecosystem, demonstrating the viability of sustainable practices.

2. Organic Valley Cooperative, USA:

The Organic Valley Cooperative is a successful example of an organic farming collective. Comprising numerous small-scale farmers, this cooperative emphasizes organic practices, animal welfare, and fair compensation for farmers. The cooperative model ensures the economic sustainability of member farms while meeting the growing demand for organic products.

3. Kraay's Market and Garden, Alberta, Canada:

Kraay's Market and Garden is a diversified farm that exemplifies sustainable and organic practices. The farm focuses on regenerative agriculture, integrating animal husbandry with vegetable production. By using animal manure to fertilize crops and rotating animals through different areas of the farm, Kraay demonstrates a holistic approach to sustainable farming.

A Comprehensive Exploration of Sustainable and Organic Practices provides an in-depth understanding of the principles, certification processes, and successful case studies within the realm of sustainable and organic animal husbandry.

Embracing these practices not only addresses contemporary concerns about environmental impact and animal welfare but also fosters a resilient and ethical approach to producing animal products. As consumers increasingly prioritize sustainability and organic choices, this book serves as a valuable resource for both producers and enthusiasts seeking to contribute to a more environmentally conscious and humane future in animal husbandry.

Emerging Trends and Technologies

In the rapidly evolving landscape of animal husbandry, staying abreast of emerging trends and technologies is critical for sustainable and efficient agricultural practices

Precision Farming and Technology Applications in Animal Husbandry:

1. Precision Livestock Farming (PLF):

Precision farming techniques have revolutionized the way animals are raised and managed. PLF involves the use of advanced technologies such as sensors, GPS, and data analytics to monitor and manage individual animals or herds. This allows farmers to optimize feeding, monitor health, and enhance overall productivity.

2. IoT and Sensors:

The integration of Internet of Things (IoT) devices and sensors in animal husbandry provides real-time data on various aspects of animal health and behavior. Wearable sensors can monitor vital signs, and feeding patterns, and even detect early signs of diseases, enabling prompt intervention and improving overall herd management.

3. Automated Feeding Systems:

Automated feeding systems are becoming increasingly prevalent in modern animal husbandry. These systems use technology to precisely measure and distribute feed, ensuring each animal receives the appropriate nutrition. This not only optimizes resource utilization but also enhances the health and growth of the livestock.

4. Robotics in Animal Handling:

Robotics plays a vital role in minimizing human labor while improving the efficiency of animal handling. Robotic systems can be employed for tasks such as milking, cleaning, and even monitoring animal movements. This reduces stress on both animals and farmers while increasing overall productivity.

Future Trends in Animal Husbandry:

1. Genomic Selection:

The future of animal husbandry is likely to see increased emphasis on genomic selection. By analyzing the genetic makeup of animals, farmers can make informed decisions about breeding, disease resistance, and desirable traits. This trend has the potential to significantly accelerate genetic improvement in livestock populations.

2. Vertical Farming for Animal Feed:

As concerns about sustainable agriculture grow, vertical farming is emerging as a viable solution. Growing animal feed in controlled indoor environments allows for year-round production, reduces the environmental impact of traditional crop cultivation, and ensures a consistent and high-quality feed supply for livestock.

3. Blockchain for Traceability:

The adoption of blockchain technology in animal husbandry offers enhanced traceability and transparency in the supply chain. From farm to table, blockchain can track every stage of production, providing consumers with accurate information about the origin, conditions, and handling of animal products.

4. Personalized Nutrition and Health Monitoring:

Advances in technology will enable personalized nutrition plans for individual animals based on their specific needs and health status. Monitoring tools and wearable devices will become more sophisticated, allowing farmers to tailor diets and healthcare interventions for each animal, optimizing overall well-being.

CONCLUSION

In the concluding chapters of "Animal Husbandry: A Comprehensive Exploration of Sustainable and Organic Practices" and "Animal Husbandry: Exploring Emerging Trends and Technologies," the reader is left with a profound understanding of the evolving dynamics within the realm of animal farming.

These books collectively emphasize the imperative for a holistic approach that aligns with environmental sustainability, animal welfare, and technological innovation.

In the sustainable and organic practices context, the conclusion underscores the importance of embracing ethical and environmentally conscious methods in animal husbandry. It resonates with a call to action for farmers, policymakers, and consumers to collectively contribute to a more harmonious and sustainable relationship between humans, animals, and the environment. The principles of environmental stewardship, animal welfare, biodiversity conservation, and closed-loop systems are presented not just as ideals but as pragmatic solutions for the future of agriculture.

On the other hand, the conclusion of the book exploring emerging trends and technologies in animal husbandry paints a picture of a future where precision farming, artificial intelligence, and advanced data analytics are integral components of successful animal husbandry.

It leaves the reader with a sense of excitement about the possibilities offered by technologies like IoT, robotics, and blockchain, emphasizing that these innovations are not just on the horizon but are already shaping the present and future of animal farming.

In essence, these conclusions encapsulate the transformative journey undertaken within the pages of these books, inviting readers to contemplate the path forward for a more sustainable, ethical, and technologically advanced era in animal husbandry.